CONTENTS

A NOTE TO TEACHERS AND PARENTS

This book is full of exciting activities for children and fascinating facts about stars. Every page is written directly for children, with information and activities that they can easily understand.

Topics in this book cross the curriculum: science, language arts, math, history, and art. For example, children will learn about latitude and using the stars to find directions. They will make up myths about constellations and create their own star patterns. In addition, they will conduct simple experiments and learn the importance of observation and imagination in the scientific world.

If you are working through this book as a class, photocopy the activities and give a page to each child (or each group of students, if you are doing group activities). A visit to a local planetarium makes an excellent field trip.

Many of the projects are ideal for homework assignments. Watching a meteor shower, checking the dawn sky for disappearing stars, and searching for satellites require viewing the night sky. Interviewing a star is a good research project.

Parents can also help their children with the activities in *Stars*. They can read directions aloud and do activities with their children. For example, parents can assist with making the star charts or join in on viewing meteors or star clusters.

The most important thing about *Stars* is that children will enjoy exploring the world of science. You can help make their scientific explorations and experiments a success.

STARS

······················

A SCIENCE DISCOVERY BOOK

by Pat Tompkins
illustrated by Marilynn G. Barr

Dedicated to
Elsie P. Tompkins

Publisher: Roberta Suid
Design & Production: Scott McMorrow
Cover Design: David Hale
Cover Art: Mike Artell

For a complete catalog, please write to the address
below:
P.O. Box 1680
Palo Alto, CA 94302

Call our toll-free number: 1-800-255-6049
E-mail us at: MMBooks@aol.com
Visit our Web Site:
http://www.mondaymorningbooks.com

Monday Morning Books is a registered trademark of
Monday Morning Books, Inc.

WELCOME TO THE NIGHT SKY

You have a free ticket to see a new show every night. You don't need any TV set or Internet to enjoy looking at the night sky. At first, it may look like a bunch of white dots twinkling on a dark background. However, soon you'll see patterns and notice different sizes and colors. You'll learn how to tell a star from a planet. You'll also learn the names of stars and constellations.

People who study the stars are called astronomers. To become a star scientist, start by filling out your Stellar Astronomer's Badge. Write your name, age, address, and favorite star or group of stars on the badge pattern. You might choose the Little Dipper or the North Star. Cut out the badge, and cover it with clear contact paper to protect it.

Before you continue, see how much you already know about stars by taking the Stars Quiz. Then check the Answers to the Quiz. Try testing your friends' or your family's star knowledge with the quiz.

STELLAR ASTRONOMER

Use your Stargazer's Notebook to begin recording any information you find interesting. Once you fill up this page, continue by using blank notebook paper. Staple a construction paper cover to your notebook. Decorate the cover with your own drawings or use shiny stick-on stars. Astronomers learn more about the stars all the time. Often the weather reports in daily newspapers include information about the planets and stars that are visible that night, along with the times the sun rises and sets and the phases of the moon. Learn more about stars by searching the Web and by visiting a planetarium.

What you see in the night sky depends on where you are and the time of year. If you live in a big city, streetlights and neon advertising signs brighten the night sky. This makes it more difficult to see the stars clearly. You will see more stars if you live in the country. Where you are in relation to the equator affects what you see, too.

Always check the night sky whenever you travel—at the beach or camping in the woods or visiting friends in another town. There are a trillion stars to see, so start looking!

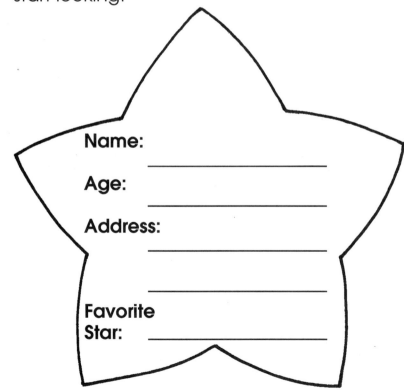

Name:
_____

Age:
_____

Address:
_____

_____

Favorite Star: _____

STARS QUIZ

1. A shooting star is a traveling star.

2. The Sun is one of the biggest stars.

3. On a clear night, away from city lights, you can see 1,500 stars, without using a telescope.

4. All stars are about the same distance from Earth.

5. The Big Dipper is the biggest constellation.

6. Scientists who study the sky are called astronomers.

7. The Milky Way is the name of our galaxy.

8. The Evening Star is one of the brightest stars.

9. People who live in Australia see the same stars as people who live in the United States.

10. The zodiac contains 12 constellations.

Gemini

Leo

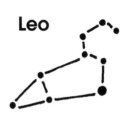

Libra

Sagittarius

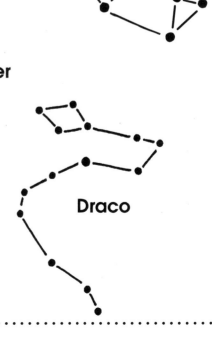

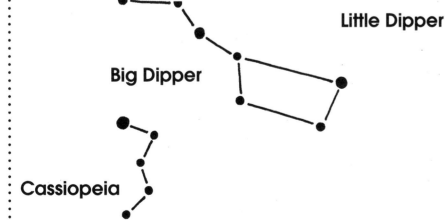

Little Dipper

Big Dipper

Cassiopeia

Draco

ANSWERS TO QUIZ

1. A shooting star is a traveling star. FALSE! A shooting star is really a meteor, not a star.

2. The Sun is one of the biggest stars. FALSE! The Sun is the star closest to our planet. That's why it looks big. It's just an average size star.

3. On a clear night, away from city lights, you can see 1,500 stars without using a telescope. TRUE! Without clouds or the light pollution of city lights, 1,500 stars are within view.

4. All stars are about the same distance from Earth. FALSE! They look like they're the same distance away, but some are much farther away than others.

5. The Big Dipper is the biggest constellation. FALSE! The Big Dipper is only part of a constellation, Ursa Major, that is one of the largest.

6. Scientists who study the sky are called astronomers. TRUE!

7. The Milky Way is the name of our galaxy. TRUE! In fact, the word galaxy comes from the ancient Greek word for milk.

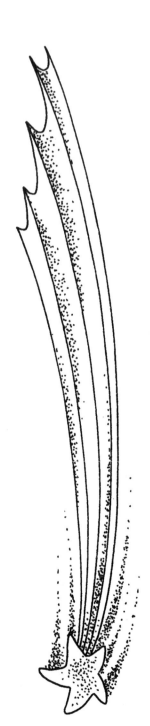

8. The Evening Star is one of the brightest stars. FALSE! It is bright, but the Evening Star is another name for Venus, a planet.

9. People who live in Australia see the same stars as people who live in the United States. FALSE! People who live south of the equator, like people in Australia, see different stars from those seen by people in the United States. The United States is north of the equator.

10. The zodiac contains 12 constellations. TRUE! Each constellation of the zodiac represents a month's worth of days.

STARGAZER'S NOTEBOOK

PLANET EARTH

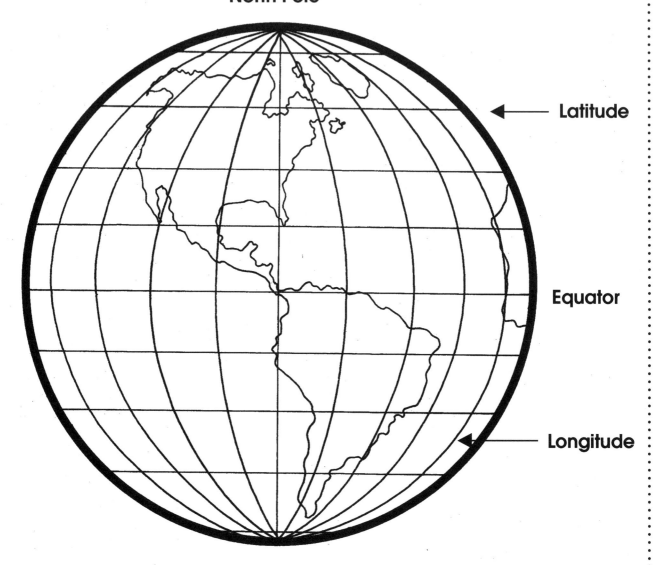

North Pole

Latitude

Equator

Longitude

South Pole

You could line up more than 100 Earths across the equator of the Sun!

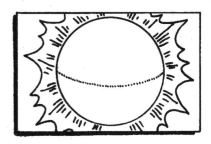

WHAT ARE STARS?

A star is a mass of gas, shining with nuclear energy. It starts when gas and dust are squeezed together tightly and become hot. The center of the star gets so hot that nuclear fusion begins. Nuclear fusion occurs after the temperature reaches millions of degrees. In fusion, small particles combine into one big particle.

That's the official, scientific definition of a star. But stars are also beautiful objects that people have always watched with interest.

Ancient people used the stars for calendars and to guide them in their travels. They made up patterns in the night sky. They gave them names to help them tell one star from another. And they told stories about those patterns. Whether they lived in Asia or Africa or on Greek or Pacific islands, people told stories about the stars.

Today, we still use the names that people gave to stars thousands of years ago. And when we want to draw a star, we usually create a figure with five points. When we do that, we're borrowing a symbol from the ancient Egyptians: ★ To them, a five-pointed symbol represented a soul. But stars are actually round. They may look like tiny jewels to us, but they are really the largest objects we see!

MEET ASTRONOMERS

Astronomers are scientists who study the stars. They also study planets, comets, meteors, and moons. The word astronomy means "arrangement of the stars."

For 5,000 years, astronomers have kept records of what they see. About 400 years ago, the telescope was invented. This let astronomers see many details: craters of the moon, sunspots, moons of other planets, and more stars.

Famous astronomers from history include Claudius Ptolemy, Nicolaus Copernicus, Tycho Brahe, and Johannes Kepler. Galileo Galilei, Isaac Newton, and Albert Einstein all made important contributions to astronomy. Amateur astronomers, people who watch the sky as a hobby, also make important discoveries.

Have you heard of the Hubble Space Telescope? It's named for Edwin Hubble. Find out who he was by looking in astronomy books or searching on the Internet.

Just for Fun: Back in Time

Imagine that you lived 400 years ago and that you were the first person to look through the telescope. Your friends and family have never seen the things you've seen. Write a description of what you saw through the telescope. Use plenty of details so that your readers can picture what you're writing about. You might describe the craters of the moon, sunspots, and moons of other planets.

TRICKY STARS

All stars are not the same. They have different colors and temperatures. Some are closer or larger than others. Stars look the same because they are so far from Earth.

Astronomers use telescopes and other tools to learn about the stars. They have found that some bright stars are really two stars together. Astronomers first discovered that the Earth orbits the Sun even though the Sun looks like it moves across the sky.

You can't always be sure of what you see. Hold a pencil straight up and down in front of your face. Close one eye, then quickly open it and close your other eye. Do this several times. The pencil appears to jump from side to side, but it hasn't moved at all.

Now try it again, but look past the pencil to something on the other side of the room, like a lamp. You'll notice that the object that's far away doesn't jump, but the pencil does.

Scientists call this parallax. It means that close stars seem to move while those farther away seem fixed.

Just for Fun: A Trick of the Light

Here's another example of how you can't always be sure of what you think you see. Take your pencil (or a drinking straw) and place it in a glass half full of water. The pencil looks bent, but you know it's straight. The pencil doesn't really bend. Rays of light bend.

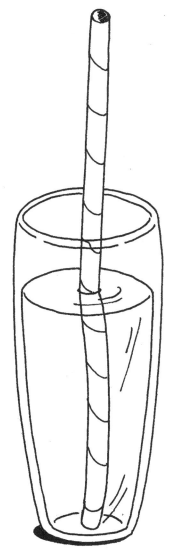

STAR PATTERN

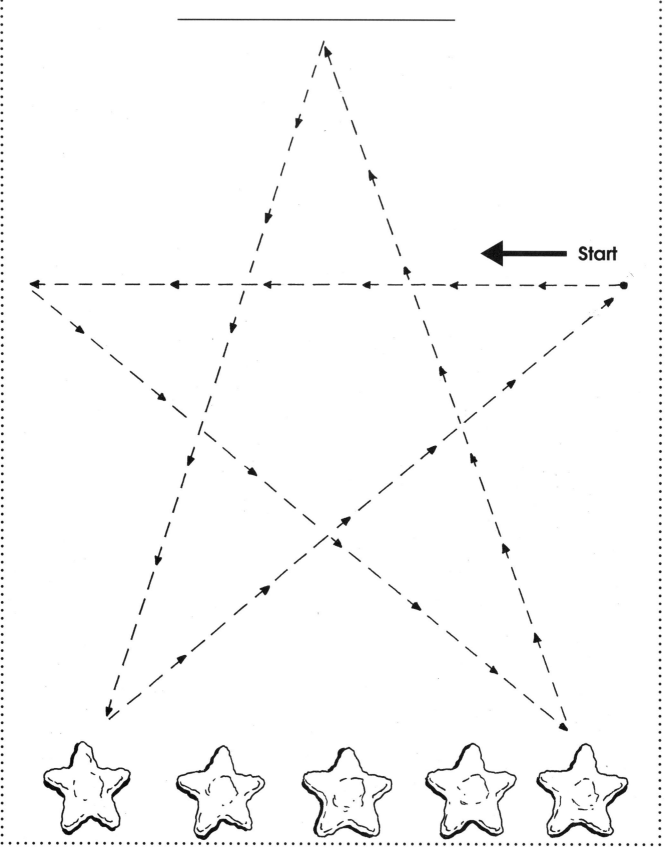

Start

Stars 1999 Monday Morning Books, Inc.

NAMING THE STARS

Over the centuries, astronomers in different countries have named the stars. Some star names are from the Greek—Procyon and Sirius. Some are Latin—Polaris, Regulus, and Spica. However, most of the names are Arabic.

Betelgeuse (pronounced BETel juice) sounds like a funny name. In Arabic it was *yad al-gawza*, which means "the shoulder of the giant." Why? Because it refers to a very bright star in the large constellation Orion, the big hunter. However, when Europeans borrowed Arabic names, sometimes the spelling changed. Below are names of some of the brightest stars:

Name	Language	Meaning
Altair	Arabic	The flier
Sirius	Greek	Glowing
Spica	Latin	Spike of grain
Aldebaran	Arabic	The follower
Arcturus	Greek	Bear watcher
Vega	Arabic	The falling
Rigel	Arabic	Foot
Mirfak	Arabic	Elbow

Just for Fun: Word Wonders

English has more words than any other language! It borrows words from many languages. Use a dictionary to find the meaning of some names of constellations, which are groups of stars. The dictionary tells the language the word is from and what it first meant. Start with Cygnus, Leo, Ursa Major, Aries, and Taurus. Add them and other interesting words to your Stargazer's Notebook.

COLORS OF STARS

The Sun usually looks yellow, orange, or red. The Sun is the star closest to our planet Earth, so it looks larger than the white stars at night.

Are stars white? Some stars are. Others are orange, red, blue—even brown and black. If you see an object in the night sky that's green or blue-green, it's probably a planet, not a star. (If you see an object that has a red blinking light, it's probably an airplane!)

At first glance, the night sky appears to be black and white. The different colors of stars are easiest to see with the brightest stars. Arcturus is yellow-orange. Rigel is blue-white. Betelgeuse is red-orange. The colors show how hot the stars are. Blue stars have the highest temperature.

How hot is a star? Very, very hot. The surface temperature of the Sun, for example, is 11,000 degrees F (6,100 degrees C). And the surface is the coolest part. It's like a bowl of soup. The soup nearest the edge of the bowl is cooler than the soup in the middle.

Just for Fun: Star-Studded T-Shirt

Paint a T-shirt with different colored stars. Use red, yellow, and blue fabric paint. First, place newspaper inside a white T-shirt so the paint won't seep through. Write a message on the front of the shirt: **I'm a Hot Star** or **I'm a Cool Star**. Let the shirt dry, and then wear it.

HOW FAR IS THAT STAR?

Stars seem to be the same size and distance from the Earth. However, some are closer than others. After the Sun, the closest stars to Earth are Proxima Centauri and Alpha Centauri. They are just over four light-years away. That's more than 25 trillion miles (40 trillion km). Yet you can see stars that are 1,000 times farther away—without a telescope.

Astronomers use "light-years" to describe huge distances. A light-year measures distance, not time. It refers to how far light travels in a year. One light-year is 5.8 trillion miles (9.5 trillion km). Light travels at 186,282 miles (300,000 km) per second. That means light can circle the Earth seven times in one second!

The brightest stars aren't always the closest. Rigel is one of the ten brightest stars. But it is 900 light-years away. Compare this to Sirius, the Dog Star. It's less than nine light-years away, but it's not 100 times brighter than Rigel. This means Rigel is very big. It's called a supergiant star! Our Sun looks bright because it is just eight light-minutes away.

Just for Fun: Flashlight Star

You can see how distance, size, and brightness are related. Go into a dark room with a flashlight. Stand a few feet in front of a blank wall. Shine the flashlight on the wall. Then step back several feet. Point the flashlight at the wall again. Notice how the brightness changes as you move away from the wall. What happens to the size of the circle of light?

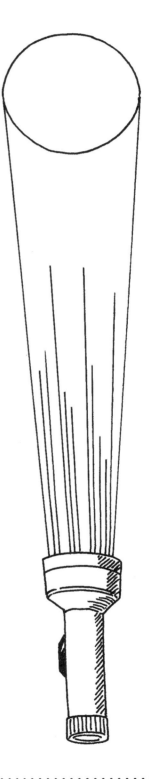

THE SUN IS A STAR

The Sun is not the brightest star, but it is the closest. It is 93 million miles from Earth. That's why it looks so big compared to stars. It takes eight minutes for light to travel from the Sun to our planet. Other starlight takes years to reach us.

The Sun looks circular. In fact, all stars are round, not pointed! It is only an average-size star, but appears very bright to us. It's so bright that people wear sunglasses to protect their eyes. **Never** look directly at the Sun!

Like other stars, the Sun is made up mostly of hydrogen gas heated to form helium. The surface is 11,000 degrees F (6,100 degrees C)! The hotter core is millions of degrees.

Until about 400 years ago, most people believed the Sun moved around the Earth. Actually, the Earth orbits the Sun and depends on it in many ways. The Sun tells us what time it is, when to plant and harvest crops, and when to get up and go to sleep. It is our source of energy.

Some ancient people thought that when the Sun set each evening, it exploded into little pieces. These pieces of Sun created the stars of the night. Every dawn was the birth of a new Sun.

Just for Fun: Dawn Patrol

Get up early one morning to watch the stars fade during the hour before dawn. The stars seem to disappear, but they don't go anywhere. The Sun is so bright that we can't see the other stars during the day.

THE DOG STAR

The Dog Star is the brightest star in the night sky. It is part of the constellation called Canis Major, the Great Dog. The official name of the Dog Star is Sirius, which means "sparkling" or "glowing" in Greek. The Dog Star is two-and-a-half times bigger than the Sun. It looks smaller because it's farther away. (Even so, Sirius is one of the closest stars, less than nine light-years away.)

For ancient astronomers, the Dog Star was closest to the Sun in summer and appeared during the hottest months. That's where we get the phrase "dog days" for hot weather. Today, Sirius is brightest in the winter sky.

The Dog Star has a companion star, Sirius B, known as the Pup. Even with the help of a telescope, it is hard to see the pair as two stars. Although the Dog Star is much brighter than the Pup, they are close together—only about one billion miles apart.

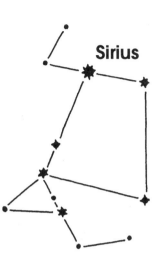

Sirius

Just for Fun: Constellation Cow

Most of the constellations in the zodiac are animals: the goat, crab, ram, fish, bull, and lion, for example. There are also bird constellations: the eagle, swan, and crow. There are horses and bears, plus a dragon, serpent, and whale. However, there's no cat constellation. Pick an animal you like that isn't a constellation and create one of your own: a kangaroo, a cow, a bat, or a cat.

THE NORTH STAR

Aside from the Sun, the North Star is the most useful star. It barely makes the Top 50 list of the brightest stars, but it shows where north is. Unlike all the other stars moving across the sky, it stays put.

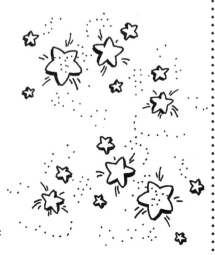

The North Star is easy to find. It's the last star in the handle of the Little Dipper. The two stars farthest from the handle of the Big Dipper also point toward the North Star. Draw an imaginary line between these two stars, then extend the line so it's about seven times longer, and you'll reach the North Star. However, if you live south of the equator—in Argentina or Australia, for example—you won't be able to see this star.

As the Earth spins in orbit around the Sun, its axis points at the North Star. All the other stars seem to revolve around it.

Travelers have long used the North Star to help them find their way. It's also known as the Pole Star, Ship Star, Polaris, and Lodestar. In India, it is called the "Pivot Point of the Planets."

Just for Fun: Locating Latitude

You can also find the North Star by finding your latitude. Look at an atlas or globe. Latitude refers to the imaginary lines that circle the Earth from east to west. If you live in New Orleans, you are near the 30th latitude. The North Star would be 30 degrees above the horizon, or about one third of the way up from the earth. The closer you are to the North Pole, the higher in the sky the North Star appears.

THE BIG DIPPER

The Big Dipper is the one constellation everyone can find. However, it's not really a constellation. The seven stars that form the Big Dipper are what astronomers call an asterism. An asterism is a group of stars that are part of a larger group. In this case, the Big Dipper is part of a constellation called Ursa Major (which means the Big Bear in Latin). It is a big constellation, but it doesn't look much like a bear.

Because the Big Dipper is easy to find, it's useful as a pointer to other stars. One end points to the North Star. The other end, the curved handle, leads in an arc to Arcturus. This is one of the brightest stars in the northern night sky.

Look carefully at the middle star in the handle of the Big Dipper. You'll see that it's actually two stars. These are the clearest double, or binary, stars. The brighter star of the pair is Mizar. The other is called Alcor.

Just for Fun: Name Game

Around the world, people have seen different shapes in the stars of the Big Dipper and of Ursa Major. This group of stars has been called the Chariot, the Plough, the Wagon, the Grain Measure, and the Reindeer. What do these stars look like to you? Create your own name for the stars of the dipper or the bear.

Ursa Major/Big Dipper

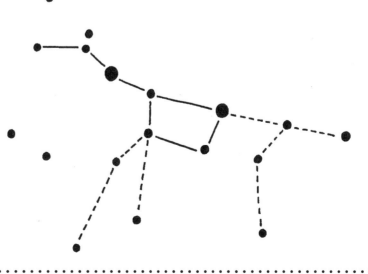

THE LITTLE DIPPER

Like the Big Dipper, the Little Dipper is visible all year in the northern hemisphere. It also has seven stars. It is smaller and dimmer than the Big Dipper. Another name for the Little Dipper is Ursa Minor (the Little Bear).

Once you find the Big Dipper, you can find the Little Dipper. That's because the Big Dipper points to the North Star. And the last star in the handle of the Little Dipper is the North Star. You'll always find the two dippers on opposite sides of the North Star.

The two dippers don't resemble bears, which have very short tails. Yet people in ancient Greece and various Indian tribes in North America called them bears. Although these people lived on opposite sides of the world and had no contact with each other, they both saw bears in the night sky.

Besides naming groups of stars as animals or other figures, ancient people made up stories about them. Many myths about Ursa Major and Minor are based on the way these constellations circle around the North Star. In some stories, the bears are trying to escape from hunters and dogs chasing them.

Just for Fun: Star Story

Create a story about the stars in the Big and Little Dippers. You don't have to use dippers or bears. For example, you could tell a story about a Mother Skunk playing tag with her Baby Skunk. (The Sioux Indians saw a skunk instead of a bear.) Be sure to give your characters interesting names.

DiPPERS MAP

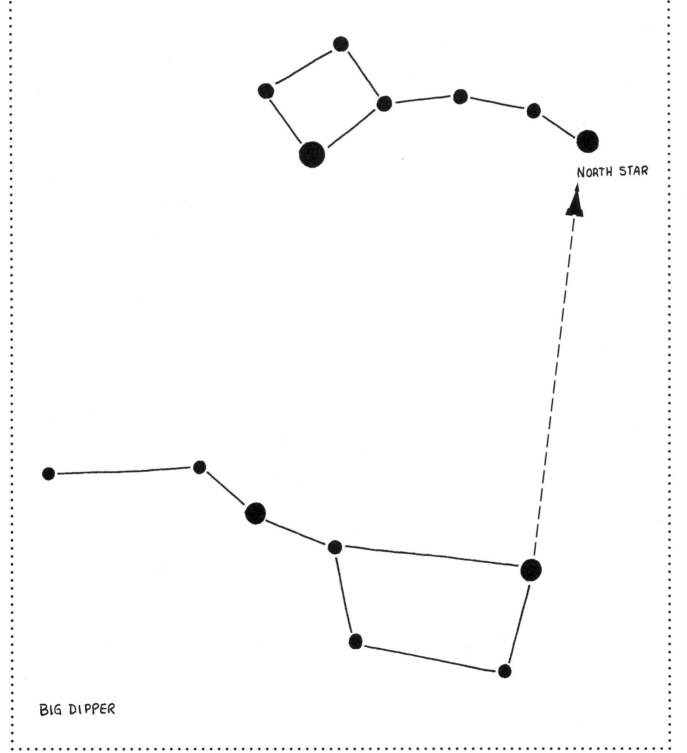

LITTLE DIPPER

NORTH STAR

BIG DIPPER

CONSTELLATIONS

Constellations are groups of stars that form a pattern. People around the world have seen different pictures in the night sky. They first used these patterns to keep track of time. Before people figured out writing systems, the stars were their calendar!

Astronomers have agreed on 88 different constellations. These cover all of the sky. In the Northern hemisphere, we can see 37 of the constellations. South of the equator, 51 of them are visible. The oldest and best-known constellations are those north of the equator.

Most constellations represent people or animals. A few are named for objects, such as Libra, a scale for weighing. Many of the constellations don't look anything like their name. However, some look like the items, animals, or people they're named for. Leo, for example, has a curved head that looks like a lion's head.

Just for Fun: Create Constellations

Create your own constellation. For example, the stars that make up the constellation Sagittarius are supposed to resemble an archer. However, many people see a teapot instead. Perhaps the curved tail of Scorpius looks like a fish hook to you. Or you may see a kite or a pair of pliers or a hat somewhere above. Draw the object you see and write a few lines telling about it.

CONSTELLATION CREATION

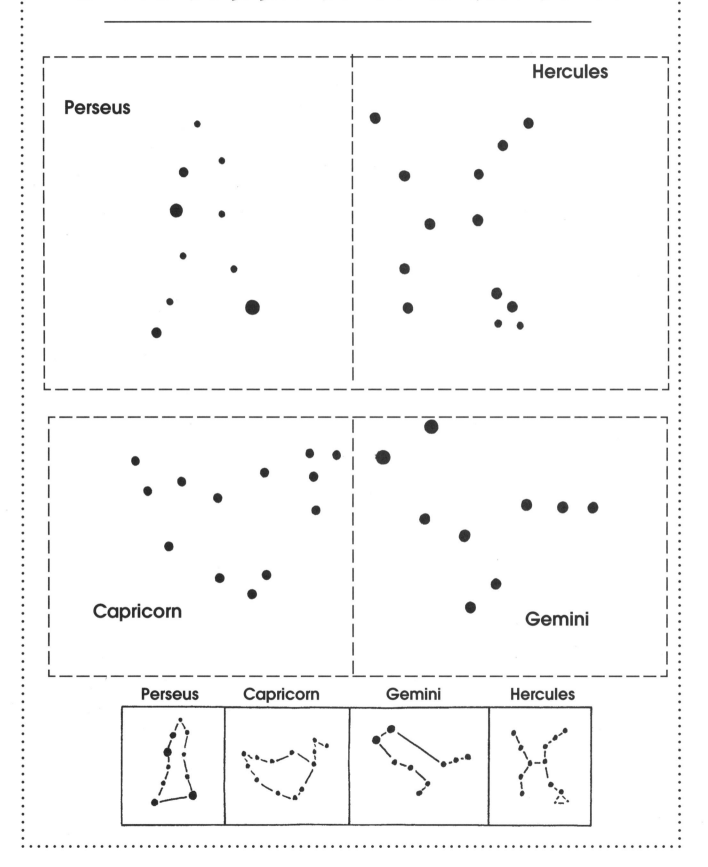

Perseus

Hercules

Capricorn

Gemini

| Perseus | Capricorn | Gemini | Hercules |

ORION THE HUNTER

In Greek mythology, Orion was a hunter who boasted that he could kill any animal. Instead, he was killed by a scorpion. The constellation Orion is one of the easiest constellations to find. It contains two very bright stars. Also, it looks like what it is named for, a giant hunter. Betelgeuse, which marks a shoulder, and Rigel, which marks a foot, are among the ten brightest stars.

Besides brilliant stars, Orion contains two famous nebulas. A nebula is the Latin word for cloud. It looks like a smudge and is made of gas and dust. One of the few nebulas we can see without a telescope is the Great Nebula. It is found in Orion's sword, which hangs from the three stars of his belt.

Below the lowest of the three belt stars is the Horsehead Nebula. It's best seen through a telescope. Other stars light up the outline of a horse's head.

Just for Fun: Follow the Hunter

Once you've located Orion, you can use it to find other major stars. Like the Big Dipper, Orion points toward several bright stars. One is Sirius, the Dog Star. Another is Procyon in Canis Minor. (These two "dogs" accompany the hunter.)

After you find Orion, draw your own picture of the hunter. You can use sequins, buttons, or small plastic "jewels" for the different stars of this famous constellation.

Orion

Tammy Reed

ORION

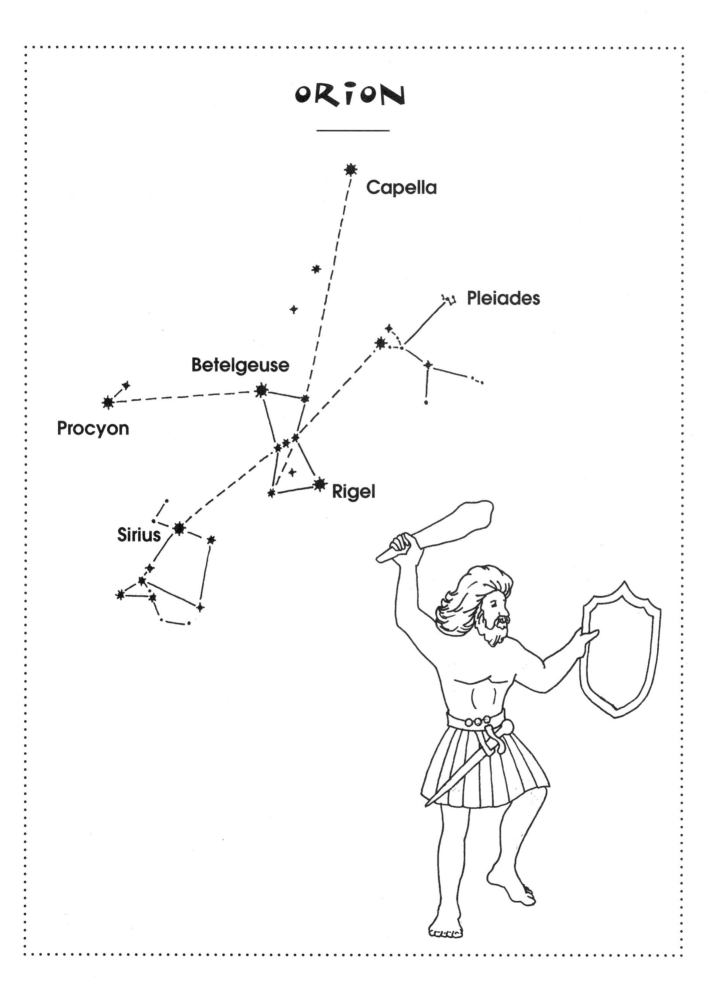

Capella

Pleiades

Betelgeuse

Procyon

Rigel

Sirius

TALKING WITH A BIG STAR

• *You certainly have an unusual name, Betelgeuse. What does it mean?*

It's an Arabic name. As you know, I'm part of the constellation Orion. But the Arabs saw a sheep instead of a hunter. So my name originally meant "armpit of the sheep." Yuck, huh? Nowadays, I'm usually known as the "shoulder of the hunter."

• *What type of star are you?*

I'm a red supergiant. And I mean super.

• *How big are you?*

I'm the biggest thing you can see.

• *The biggest star?*

The biggest single object in the universe that you folks on Earth can see. I'm much bigger than a planet like Jupiter.

• *You don't look that big.*

That's because I'm 500 light-years away. I'm almost 300 times bigger than your Sun. I give off 17,000 times more light than the Sun does. That's big.

• *Even though you're so far away, is it true that you're one of the brightest stars?*

Yes. Only a few stars are brighter than me. And I have the coolest color, orange-yellow.

• *I've noticed that sometimes you look brighter than at other times. It's like your size changes. Do you have a weight problem?*

No. All red supergiants vary in size. I sort of shrink for a while and then I puff up. Then I shrink again. But I'm always big.

• *Thanks for talking with us today, Betelgeuse.*

My pleasure.

Just for Fun: Become a Reporter

Choose a star you want to know more about. Before interviewing your star, use books to find out more about it. That way, you'll be able to ask good questions. Write your interview for the Star News. You can add other space-related stories, poems, or songs to your Star News, as well.

STAR★NEWS

THE ZODIAC

The zodiac is an ancient system for keeping track of the seasons. There are 12 constellations in the zodiac, one for each month. During the year, the Sun and planets appear to travel past these 12 constellations. (The movement is really from the Earth's orbit of the Sun.)

The Greek word zodiac refers to a circle of animals. Most constellations in the Greek zodiac are animals. In the Chinese zodiac, all of the constellations are animals. Instead of January 1, the zodiac starts on March 20 just before the Sun crosses the equator on the first day of spring.

Greek Zodiac	Chinese Zodiac	Time of Year
Aries (ram)	Rat	March 20-April 18
Taurus (bull)	Ox	April 19-May 19
Gemini (twins)	Tiger	May 20-June 20
Cancer (crab)	Hare	June 21-July 21
Leo (lion)	Dragon	July 22-August 21
Virgo (virgin)	Serpent	Aug. 22-Sept. 21
Libra (scales)	Horse	Sept. 22-Oct. 22
Scorpio (scorpion)	Ram	Oct. 23-Nov. 21
Sagittarius (archer)	Monkey	Nov. 22-Dec. 20
Capricorn (goat)	Rooster	Dec. 21-Jan. 18
Aquarius (water bearer)	Dog	Jan. 19-Feb. 17
Pisces (fish)	Pig	Feb. 18-March 19

Just for Fun: Find Your Sign

What's your "sign"? It depends on your birthday. If you were born on October 28, you are a Scorpio. Find out what "sign" a friend is by looking at the list above. Then use the art on the Zodiac Cards to make a birthday card for that friend. Decorate the card with sequins or glitter.

ZODIAC CARDS

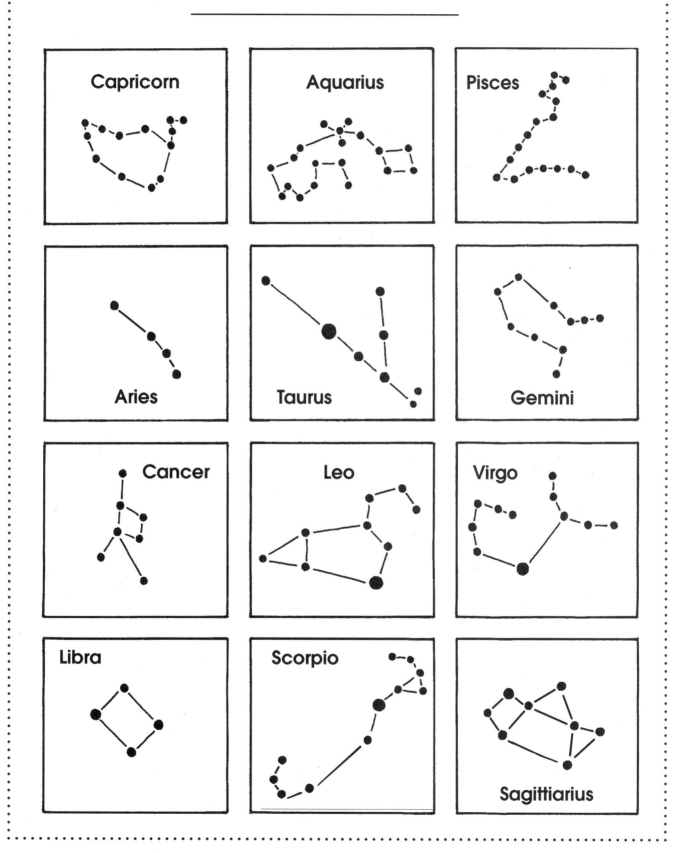

Capricorn

Aquarius

Pisces

Aries

Taurus

Gemini

Cancer

Leo

Virgo

Libra

Scorpio

Sagittiarius

SISTERS, CAMELS, AND PIGLETS

Not all groups of stars are constellations. However, they have names and stories even though they don't form pictures. They are called star clusters. Two famous clusters are the Pleiades and the Hyades.

Both the ancient Chinese and Greeks called the Pleiades the Seven Sisters. The Arabs named them the Little Camels. They're found near the V-shaped head of the constellation Taurus. On your own, you'll be able to see at least six of the seven sisters. Through binoculars, you can see about 50 stars.

The Pleiades contain hundreds of stars. The brightest are hot blue giants. These are young stars. Dinosaurs would not have seen the Pleiades. The Hyades are five stepsisters of the nearby Pleiades. In Greek, Hyades means piglets.

Use binoculars to look at star clusters. In the Northern Hemisphere, the Pleiades and the Hyades appear in the winter night sky. When you find a star cluster, draw your own version of it in your Stargazer's Notebook.

Just for Fun: Little Camels

Make a paper necklace or belt using the camel pattern. Trace it onto construction paper and cut it out. Punch holes in either side of each pattern and string the camels together to make a caravan. Glue on sequins or star stickers for decoration.

FAMOUS STARS TRADING CARDS

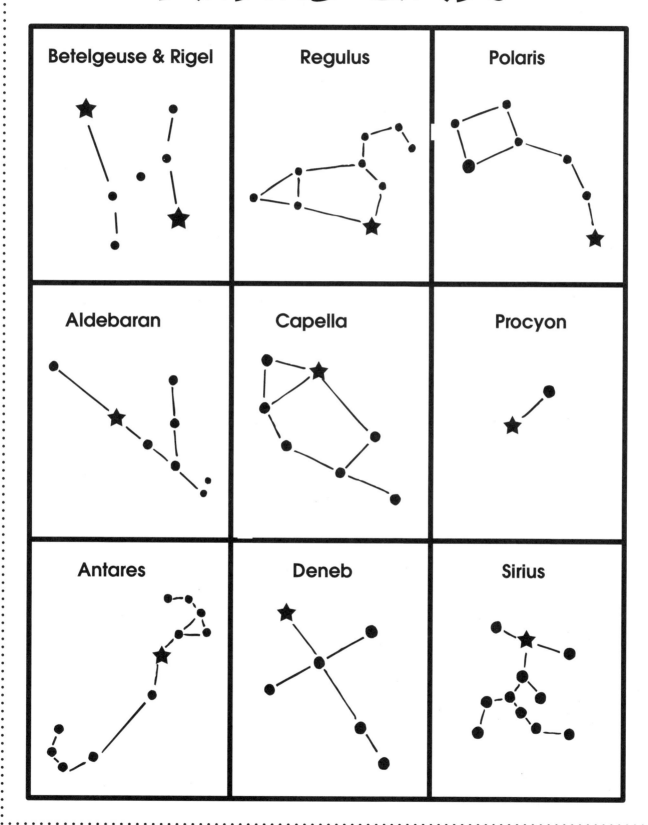

Betelgeuse & Rigel	Regulus	Polaris
Aldebaran	Capella	Procyon
Antares	Deneb	Sirius

FAMOUS STARS
TRADING CARDS

(Ursa Minor) Polaris, also called the North Star, is the last star in the Little Dipper's handle. It shows where North is.

(Leo) Regulus is the bright star marking the heart of the lion. Look for it in spring.

(Orion) Rigel and Betelgeuse shine brightest in the constellation of the giant hunter.

(Canis Minor) Procyon is one of the nearest bright stars, about 11 light-years away.

(Auriga) Capella is the fifth brightest star of all. Its name is Latin for "little female goat."

(Taurus) The red star Aldebaran creates a glaring eye in the face of the bull.

(Canis Major) Sirius is the fifth closest star and the brightest in the night sky. It's also called the Dog Star.

(Cygnus) Deneb marks the tail of the swan constellation. It's 1,600 light-years away, the most distant bright star.

(Scorpius) Look for Antares, a red supergiant, in the summer. It marks the heart of the scorpion.

WHERE ARE YOU?

The stars were the first maps for ancient travelers. Arabs crossing sand dunes used the stars to find their way. So did sailors far from land. By studying the constellations and the brightest stars, Polynesians sailed from Tahiti to Hawaii. That's more than 3,000 miles (4,800 km)!

For the South Pacific sailors, the three stars that we call Orion's belt were "the three canoe paddlers." When they couldn't see the North Star, they used the Southern Cross constellation, which they called Koloa, or "the duck." Without any instruments, these navigators found tiny islands in the huge ocean. The stars were their guide.

Imagine sailing across an ocean. The weather is stormy. For several nights, you can't see the stars. What could you use to help you find your direction? Write a story about what you might do.

Just for Fun: Be a Star Guide

Hobbies, such as star-gazing, are fun to share with friends. Once you know how to find different star clusters, have a party. Invite your friends over and take them on a journey through the night sky. Point out different constellations and clusters!

SHOOTING STARS

Any night, you might see a dramatic flash of light across the sky. It looks like a star falling to Earth. However, it's really not a star. It's a meteor, a bit of debris from a comet's tail.

Although a meteor looks like a star, it's tiny. Meteors usually aren't any bigger than pebbles. Some are as small as grains of sand! However, when they pass through the Earth's atmosphere, we see a streak of light. Their movement produces friction, heating meteors so that they glow. (You can make friction by rubbing your palms together. When you do that fast, your hands get warm.) Some shooting stars are orange, green, or yellow. A very bright meteor is called a fireball.

Most nights, you might see five or six meteors in an hour. You can see 50 or 60 an hour during a meteor shower! They are called showers because the meteors seem to rain down on the Earth. Here are the major meteor showers each year:

Quadrantids	January 3
Eta Aquarids	May 3
Perseids	August 12
Orionids	October 21
Leonids	November 16
Geminids	December 13

Just for Fun: See a Shower

Watch a meteor shower. Give your eyes 10 or 15 minutes to adjust to the dark. You can record your meteor sightings in your Stargazer's Notebook! Draw a picture of what you see.

STAR OR PLANET?

Because stars and planets are so far away, they look similar to us. However, they are very different. There's one easy way to tell if a bright light in the sky is a star or a planet. Stars twinkle and planets don't.

The word planet means "wanderer." The ancient Greeks saw that some objects moved around in the sky. So they called the planets "wandering stars." The others were "fixed stars" because they seemed to stay still. Venus often shines brightly soon after sunset. It's known as the "evening star" even though it's a planet. Watch it, and you'll see that it has a steady glow.

Planets are bodies that revolve around a star. They reflect starlight. Stars create their own light. They twinkle because their light passes through the Earth's atmosphere. If you were an astronaut on a spaceship, you would see stars shining but not twinkling.

Just for Fun: Sing a Song

Make up a song about the planets based on this familiar song about the stars:

Twinkle, twinkle little star
How I wonder what you are.
Up above the world so high
Like a diamond in the sky.
Twinkle, twinkle, little star
How I wonder what you are.

You could start with: "Shine on, shine on, planet bright." Or "Glowing, glowing in the sky." Share your song with your family.

STARLIKE, STAR BRIGHT

Besides planets, other objects in the night sky resemble stars. Comets look like a blur of stars with a tail. These tails are millions of miles long! The ancient Greeks called them "long-haired stars." Comets are cold, not hot. They are masses of ice and frozen gases. Some orbit the Sun in regular patterns at thousands of miles per hour. They look slow because they are so far away. The most famous comet is Halley's comet. It returns every 76 years. The next time will be in 2061.

Asteroids are another type of object in space. The Greeks called them "starlike." (Aster is Greek for star.) But asteroids are more like little planets. They are big rocks. Many asteroids orbit the Sun between Mars and Jupiter.

You can also see two types of satellites at night. Satellite comes from the Latin word for "attendant." That's because satellites are bodies that orbit a larger body. The moon is a satellite of our planet. Manmade spacecraft are another type of satellite. They carry instruments to collect information about the Earth and space.

Just for Fun: Satellite Search

Spend a few evenings searching for satellites. About 300 are big enough and low enough to see easily. The first hours after sunset are the best time to look. If you see something moving with regularly blinking lights, it's probably an airplane. Here's another clue to help you tell whether that moving light is a satellite or a plane: Satellites never travel from east to west. Record your findings in your Stargazer's Notebook.

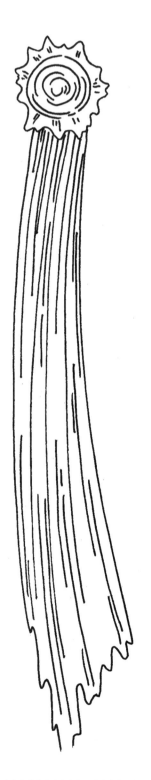

THE MiLKY WAY

A galaxy is a huge collection of stars and other celestial bodies. Our galaxy is called the Milky Way. We live near the edge of the Milky Way, about 30,000 light-years away from its center. So we can see parts of the rest of the galaxy. The best time to view the Milky Way is during the summer, far from a city or other sources of lights.

The Milky Way looks like a spill of milk in a long ribbon or river across the night sky. The white haze comes from the many stars in the Milky Way: 100 billion!

From a greater distance, our galaxy looks like a spiral, or a pinwheel, with a bulge in the middle. In the 1920s, astronomers used powerful telescopes to discover that the Milky Way is not the only galaxy. It is just one of billions.

Scientists think the universe is getting bigger all the time.

Just for Fun: Blow Up a Balloon

Here's a way to see how astronomers think the universe is expanding. Take a balloon and blow it up so that it's about the size of an orange. Make it polka dotted: mark 12 or 15 random dots using a felt-tipped pen. Then continue blowing up the balloon to its full size. You'll see that as the balloon gets bigger, the dots move farther away from each other. That's what the galaxies are doing. You can also do this activity with Silly Putty. Make several dots on a piece of Silly Putty. Then stretch it!

RED GiANTS AND WHiTE DWARVES

Stars can live for millions of years. As they age, their size and color change. Our Sun is a yellow dwarf, the most common type of star. Astronomers think it will live for 10 billion years.

Stars shine by turning hydrogen into helium. They are big balls of energy. When a star uses up its hydrogen, it swells and cools off. It becomes a red giant, like Aldebaran. A red giant can be 300 to 400 times bigger than the Sun. Betelgeuse is so big that it's called a red supergiant!

Later, the star consumes all of its fuel and it shrinks. It becomes a white dwarf, very dense and small. White dwarves are about the size of the Earth. Such small stars are hard to see.

Just for Fun: Star Fairy Tales

Now that you know about red giants and white dwarves, write a fairy tale about them. You can read fairy tales that feature giants or dwarves first, then write your own. Consider checking out a version of Snow White and the Seven Dwarves or Jack and the Beanstalk. Then rewrite the story, replacing the characters with white dwarves or red giants. Be sure to include factual information about stars in your story! Illustrate it and share it with your friends or family.

NOVAS AND SUPERNOVAS

Before telescopes, astronomers sometimes saw a strange bright light in the sky. Thinking it was a new star, they called it nova. A nova is a star that suddenly brightens as much as 1,000 times. Actually, a nova is an old star, not a new one. Scientists think a nova occurs when a white dwarf in a pair takes matter from its companion star.

A supernova is much more dramatic and rare. When a star uses up all of its fuel, it explodes. This explosion looks like a huge new star for several weeks. It can last for about a year. In one second, a supernova releases as much energy as our Sun does in 60 years!

Since the invention of the telescope, no one has seen a supernova in our galaxy. However, using telescopes, we can see the remains of a supernova that exploded in 1054. It looks like a cloud of gas in the constellation Taurus. It's called the Crab Nebula. After its supernova stage of life, a star becomes a black hole or a tiny dense neutron star.

Just for Fun: Nebula Art

Look for a book about nebulas at the library. Some books have amazing photographs of the Crab Nebula and the Horsehead Nebula. Try to see why these nebulas are named after these animals. Then draw your own pictures of nebulas. Different colored chalk on black construction paper makes interesting pictures.

BLACK HOLES AND QUASARS

When a huge star gets old, it can no longer give out enough energy to avoid collapsing. The star contracts and gets heavier. This makes it shrink even more. Finally, a star may become a black hole.

A black hole is not really a hole. It's an area where matter is packed together extremely tight. If the Earth were as dense as a black hole, it would be the size of a marble. Black holes don't give off light, X-rays, or radio waves. They don't even reflect light. Because we can't see them, it's hard to be sure they exist.

A black hole is so dense and has such strong gravity that nothing can escape it. Even light, the fastest thing in the universe, is not quick enough to avoid capture!

Astronomers have recently found another strange feature of outer space called quasars. You can't see quasars without a powerful telescope because they are so far away.

Just for Fun: More Mysteries

Scientists depend on observing things to collect facts. However, some things can't be seen. When things can't be explained, they're mysteries. Can you think of things you can't see but know are there? What about things you can see but can't touch? Moonlight is one example. Shadows are another. Keep a list of mysterious things in your Stargazer's Notebook.

PLANETARIUM FUN

Visiting a planetarium is like going to a movie theater. However, instead of looking at a flat screen in front of you, you watch the curved ceiling. A projector shines lights on the dome above. This means that even in the middle of the day, you can watch a sky full of stars.

A planetarium lets you observe the night sky in ideal conditions. You're indoors and can look at stars you might otherwise not see. You get a perfect sky. No clouds or streetlights or full moon interferes with what you can see.

If you live in or near a city, you are probably close to a planetarium. Science museums and large universities often have planetariums open to the public. Some show several different programs every day. One program might focus on the current night sky's constellations. Another might explain how people use stars for navigation. Going to a planetarium is a great way to start learning more about stars!

Just for Fun: Stars at Home

Get a large sheet of dark paper and some glow-in-the-dark stars from an art supply store. Paste the stars on the paper. (You can use star stickers if you can't find glow-in-the-dark stars.) Make up your own patterns or follow the outlines of real constellations. You could even spell out your name using the stars! Tape the paper to a wall in your room. (At the foot of your bed is a good place.) Or get help taping your star chart on the ceiling above your bed.

A CLOSER LOOK: OBSERVATORIES

Observatories are buildings that house the giant telescopes that astronomers use to study the sky. Observatories look like short silos. At night, their domes open for sky watching.

Unless you live on a mountain top or in the desert, you probably don't have an observatory in your neighborhood. Observatories are usually built in places away from cities. This is because it's easier to see stars in areas without a lot of lights.

You'll find observatories scattered around the world: Canada, South Africa, Japan, Australia. In the United States, observatories are in Arizona, California, Wisconsin, and other states. The largest telescopes on Earth are the two at the Keck Observatory in Mauna Kea, Hawaii. They have mirrors 33 feet (10 meters) across.

One way you can visit an observatory is on the World Wide Web. Many of the largest observatories have Web sites. Sometimes observatories are open to visitors during the day. However, at night, the astronomers are busy working.

Just for Fun: Name the Instrument

The European Southern Observatory in Chile features four separate large telescopes and three smaller ones. The mirrors in the bigger telescopes are 27 feet (8.2 meters) across. This observatory is a group project of European scientists. They call the combined telescopes the VLT, or Very Large Telescope. Can you think of a better name?

HUBBLE TELESCOPE

The Hubble Space Telescope (HST) is a satellite the size of a bus. It orbits the Earth every 97 minutes, traveling at 17,500 mph (29,000 kph). As the first telescope above the Earth's atmosphere, HST views the stars without clouds or gas and dust getting in the way. Astronomers hope Hubble will help them figure out how big and how old the universe is.

After Hubble was launched in April 1990, several problems appeared. The biggest problem was actually a tiny one. The main mirror had a flaw that caused blurry images. The mirror was 10-thousandths of an inch (2 millionths of a meter) too flat!

NASA trained astronauts for a year to fix Hubble. In December 1993, spacewalking astronauts made the very precise repairs. Now the pictures that Hubble sends back to Earth are very clear.

Just for Fun: Bus Ride in the Sky

The Hubble telescope is as big as a bus. It travels through space, circling at 350 miles (600 km) above our planet. Imagine that you were riding a bus through space. What would you see? What kind of things would you want to tell your friends when you got back? Write and illustrate a short story about your bus ride in the sky.

HUBBLE

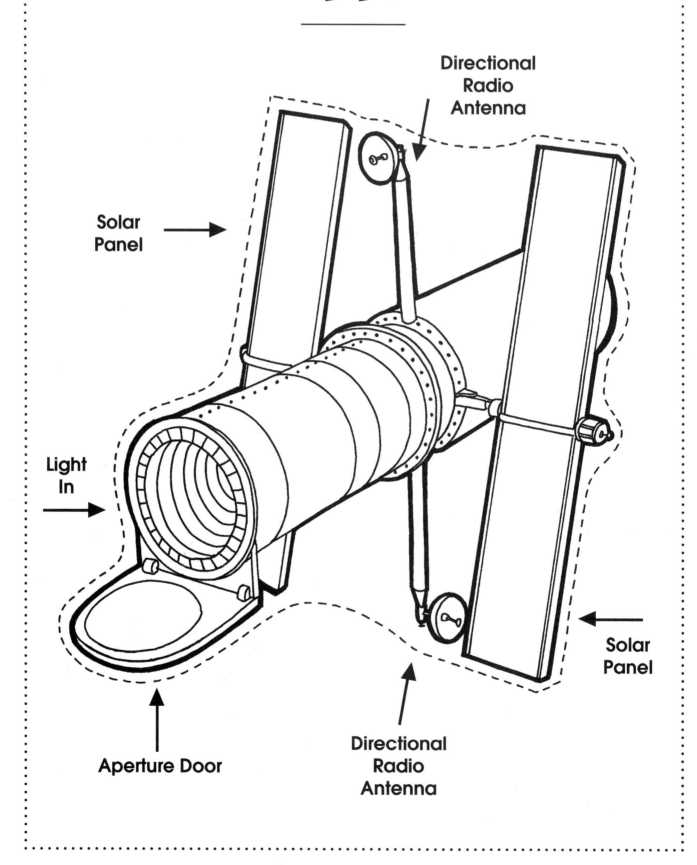

Directional Radio Antenna

Solar Panel

Light In

Aperture Door

Directional Radio Antenna

Solar Panel

STAR WORDS: A GLOSSARY

Astronomer: A scientist who studies the stars and other bodies in the universe.

Black hole: A very dense, large star that has burned out. No light escapes from black holes.

Comet: An orbiting mass of ice and frozen gas with a tail millions of miles long.

Constellation: A group of stars that resembles an imaginary object. There are 88 constellations.

Galaxy: A huge group of stars, gas, and dust that gravity holds together. Our galaxy is the Milky Way.

Light-year: How far light travels in a year, about 5.9 trillion miles (10 trillion kilometers).

Nebula: A cloud of gas and dust in space.

Quasar: A very distant small galaxy that may produce energy from a black hole in its center. Quasar means "quasi-stellar," like a star.

Red giant: A huge expanded star; it looks red because it is cooler than most other stars.

Shooting star: Not a real star, but a meteor, a falling piece of rock or metal.

Star: A big globe of gas that shines light. It creates energy through nuclear reactions. The Sun is the star closest to Earth. Although most stars look the same to us, they have different sizes, colors, and temperatures.

Supernova: A giant star that explodes because it has run out of fuel. The explosion creates a very bright light that shines for about a year.

Telescope: A tool made of lenses or mirrors that makes distant objects appear larger and closer. Telescopes were first used 400 years ago.

White dwarf: A very small star that radiates stored energy. Old stars that no longer generate new energy become white dwarves.

Zodiac: Twelve constellations in a zone of the sky in which most planets revolve around the Sun, as seen from Earth.

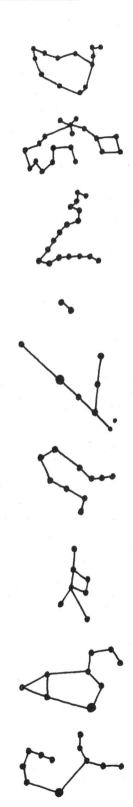

MORE ABOUT STARS

Books

Astronomy for the Beginner by Patrick Moore (Cambridge University Press, 1992).

Astronomy for Every Kid: 101 Easy Experiments That Really Work by Janice VanCleve (Wiley, 1991).

A Book of Stars for You by Franklyn M. Branley (Crowell, 1967).

The Constellations: How They Came to Be by Roy A. Gallant (Four Winds, 1979).

The Night Sky by Robin Kerrod (Benchmark Books, 1986).

Patterns in the Sky by W. Maxwell Reed (Morrow, 1951).

Skywatching by David H. Levy (Time-Life Books, 1994).

Space: A Prentice Hall Illustrated Dictionary by Michael Pollard and Felicity Trotman (Prentice Hall, 1992).

Stars & Planets by David H. Levy (Time-Life Books, 1996).

Sun Dogs and Shooting Stars: A Skywatcher's Calendar by Franklyn M. Branley (Houghton Mifflin, 1980).

Web Sites

http://www.hpcc.astro.washington.edu/scied/astro/astroindex.html

AstroEd from the University of Washington, resources for learning about astronomy

http://www.lochness.com/pltweb/pltweb.html

Planetarium Web sites from Loch Ness Productions

http://www.nasa.gov

Homepage for the National Aeronautics and Space Administration

http://www.skypub.com

SKY Online from publisher of *Sky & Telescope* magazine; search SkyLinks for many astronomy sites